T0223779

Value Rational Engineering

Synthesis Lectures on Engineering

Each book in the series is written by a well known expert in the field. Most titles cover subjects such as professional development, education, and study skills, as well as basic introductory undergraduate material and other topics appropriate for a broader and less technical audience. In addition, the series includes several titles written on very specific topics not covered elsewhere in the Synthesis Digital Library.

Value Rational Engineering
Shuichi Fukuda
2018

Strategic Cost Fundamentals: for Designers, Engineers, Technologists, Estimators, Project Managers, and Financial Analysts
Robert C. Creese
2018

Concise Introduction to Cement Chemistry and Manufacturing
Tadele Assefa Aragaw
2018

Data Mining and Market Intelligence: Implications for Decision Making
Mustapha Akinkunmi
2018

Empowering Professional Teaching in Engineering: Sustaining the Scholarship of Teaching
John Heywood
2018

The Human Side of Engineering
John Heywood
2017

Geometric Programming for Design Equation Development and Cost/Profit Optimization (with illustrative case study problems and solutions), Third Edition
Robert C. Creese
2016

Engineering Principles in Everyday Life for Non-Engineers
Saeed Benjamin Niku
2016

A, B, See... in 3D: A Workbook to Improve 3-D Visualization Skills
Dan G. Dimitriu
2015

The Captains of Energy: Systems Dynamics from an Energy Perspective
Vincent C. Prantil and Timothy Decker
2015

Lying by Approximation: The Truth about Finite Element Analysis
Vincent C. Prantil, Christopher Papadopoulos, and Paul D. Gessler
2013

Simplified Models for Assessing Heat and Mass Transfer in Evaporative Towers
Alessandra De Angelis, Onorio Saro, Giulio Lorenzini, Stefano D'Elia, and Marco Medici
2013

The Engineering Design Challenge: A Creative Process
Charles W. Dolan
2013

The Making of Green Engineers: Sustainable Development and the Hybrid Imagination
Andrew Jamison
2013

Crafting Your Research Future: A Guide to Successful Master's and Ph.D. Degrees in Science & Engineering
Charles X. Ling and Qiang Yang
2012

Fundamentals of Engineering Economics and Decision Analysis
David L. Whitman and Ronald E. Terry
2012

A Little Book on Teaching: A Beginner's Guide for Educators of Engineering and Applied Science
Steven F. Barrett
2012

Engineering Thermodynamics and 21st Century Energy Problems: A Textbook Companion for Student Engagement
Donna Riley
2011

MATLAB for Engineering and the Life Sciences
Joseph V. Tranquillo
2011

Systems Engineering: Building Successful Systems
Howard Eisner
2011

Fin Shape Thermal Optimization Using Bejan's Constructal Theory
Giulio Lorenzini, Simone Moretti, and Alessandra Conti
2011

Geometric Programming for Design and Cost Optimization (with illustrative case study problems and solutions), Second Edition
Robert C. Creese
2010

Survive and Thrive: A Guide for Untenured Faculty
Wendy C. Crone
2010

Geometric Programming for Design and Cost Optimization (with Illustrative Case Study Problems and Solutions)
Robert C. Creese
2009

Style and Ethics of Communication in Science and Engineering
Jay D. Humphrey and Jeffrey W. Holmes
2008

Introduction to Engineering: A Starter's Guide with Hands-On Analog Multimedia Explorations
Lina J. Karam and Naji Mounsef
2008

Introduction to Engineering: A Starter's Guide with Hands-On Digital Multimedia and Robotics Explorations
Lina J. Karam and Naji Mounsef
2008

CAD/CAM of Sculptured Surfaces on Multi-Axis NC Machine: The DG/K-Based Approach
Stephen P. Radzevich
2008

Value Rational Engineering

Shuichi Fukuda

ISBN: 978-3-031-79397-4   paperback
ISBN: 978-3-031-79398-1   ebook
ISBN: 978-3-031-79399-8   hardcover

DOI 10.1007/ 978-3-031-79398-1

A Publication in the Springer series
SYNTHESIS LECTURES ON ENGINEERING

Lecture #33
Series ISSN
Print 1939-5221   Electronic 1939-523X

Value Rational Engineering

Shuichi Fukuda

Adviser to System Design and Management Research Institute of Keio University
Member of the Engineering Academy of Japan

SYNTHESIS LECTURES ON ENGINEERING #33

ABSTRACT

Early in the 20th century, our world was small and closed with boundaries. And, there were no appreciable changes. Therefore, we could foresee the future. These days, however, we could apply mathematical rationality and solve problems without any difficulty.

As our world began to expand rapidly and boundaries disappeared, the problem of bounded rationality emerged. Engineeres put forth tremendous effort to overcome this difficulty and succeeded in expanding the bounds of mathematical rationality, thereby establishing the "Controllable World."

However, our world continues to expand. Therefore such an approach can no longer be applied. We have no other choice than "satisficing" (Herbert A. Simon's word, Satisfy + Suffice [2]).

This expanding open world brought us frequent and extensive changes which are unpredictable and diversification and personalization of customer expectations. To cope with these situations, we need diverse knowledge and experience. Thus, to satisfy our customers, we need teamwork.

These changes of environments and situations transformed the meaning of value. It used to mean excellent functions and exact reproducibility. Now, it means how good and flexible we can be to adapt to the situations. Thus, adaptability is the value today.

Although these changes were big, and we needed to re-define value, a greater shift in engineering is now emerging. The Internet of Things (IoT) brought us the "Connected Society," where *things* are connected. *Things* include not only products, but also humans.

As changes are so frequent and extensive, only users know what is happening right now. Thus, the user in this Connected Society needs to be a playing manger—he or she should manage to control the product-human team on the pitch.

Moreover, this Connected Society will bring us another big shift in engineering. Engineering in this framework will become Social Networking, with engineering no longer developing individual products and managing team products.

The Internet works two ways between the sender and the receiver. Our engineering has ever been only one way. Thus, how we establish a social networking framework for engineering is a big challenge facing us today. This will change our engineering. Engineers are expected to develop not only products, but also such dream society.

This book discusses these issues and points out that New Horizons are emerging before us. It is hoped that this book helps readers explore and establish their own New Worlds.

KEYWORDS

value rational design, perception, intrinsic motivation, holistic approach, atmosphere, psychological flow

Contents

Preface

Engineering was created to make our dreams come true. What we need to remember is that we had dreams and that to dream is the most important thing in the world.

However, most engineering books today discuss how to solve a problem. Indeed, there are many problems we need to solve to realize our dreams. Yet, problem solving is not our final goal. Making dreams come true is. Regrettably, we do not discuss much about dreams today. We discuss only what the problems are and how we can solve them.

Making dreams come true is the value of engineering and no matter how we do it, we will be satisfied if our dreams come true. Thus, if this goal is achieved, all paths to get there must be rational. They are very much reasonable because we needed engineering to make our dreams come true. All's well that ends well. This is the core of Value Rational Engineering.

This book, therefore, does not tell you how to solve the problems, but tells you how you can dream a good dream and how you can make it come true your way.

Everybody has his or her own dream. It varies from person to person how you make it come true. I hope this book helps you find how you can have a good dream and find your way of making it come true.

When you make your dream come true your way, you will feel the sense of achievement and fulfilment and you will be truly satisfied.

Let us explore this New World of Dream Engineering together!

Shuichi Fukuda
July 2018

Acknowledgments

I would like to thank Mr. Paul Petralia, Morgan & Claypool Publishers, for the publication of this book. For the typesetting I thank Dr. C.L. Tondo and his group at T&T TechWorks, Inc., as well as Ms. Sara Kreisman and her staff at Rambling Rose Press, Inc. for their editing work.

Shuichi Fukuda

July 2018

Introduction

This book describes how value and rationality will change in the emerging Connected Society. The fundamental change in engineering is that we need to move from individual play to team play in order to cope with the frequently and extensively changing environments and situations.

We need a wide variety of team members to adapt to such changes.

How we can organize a team and how we should manage and operate the team will be discussed. Moreover, what is very important in the Connected Society are not only products but how humans are connected. The perfect team must therefore be composed of both products and humans.

The mathematical approach has been dominant in traditional engineering. However, our world is expanding quickly, and changes are so frequent and extensive that we can no longer apply these mathematical approaches. Therefore, how we can secure value and rationality is what will be discussed.

It will be shown that value and rationality are nothing other than how we, engineers, can meet the expectations of our customers and satisfy them. In short, engineering tomorrow will become expectation management.

The paths to manage expectations are very diverse. This book shows the New Horizons and what paths there may be and how we can possibly explore them and can develop our New Worlds will be described.

CHAPTER 1

Big Shift in Engineering

1.1 CLOSED WORLD TO OPEN WORLD

Yesterday, our world was small and closed. There were clear boundaries. But, our world is quickly expanding and is changing to an Open World without boundaries (Fig. 1.1).

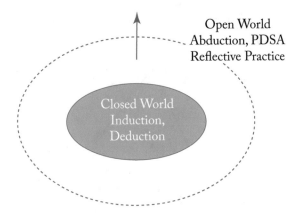

Figure 1.1: Closed World to Open World.

1.2 CHANGES OF YESTERDAY AND TODAY

Early in the 20th century, there were no appreciable changes and if there were, they were smooth, so that we could differentiate them mathematically and predict the future (Fig. 1.2).

In short, tomorrow was the continuation of today and life did not vary from person to person. Situations did not vary from time to time nor from person to person. Therefore, we could design and produce the same, or very similar, kinds of products in mass. Their expected functions were simple. We could evaluate value and apply mathematical methods easily. We could solve the problem quantitatively along the line of mathematical rationality. And, we could secure the optimum result.

But our world is rapidly expanding. Boundaries are disappearing. Now, our world is an open world. So, our life becomes very much diversified and personalized. And changes take place frequently and extensively. What makes problem solving difficult is that these changes are

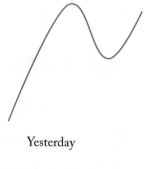

Yesterday

Smooth Change
Differentiable
Predictable

Figure 1.2: Changes of yesterday.

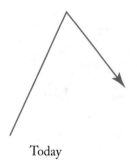

Today

Sharp Change
Not Differentiable
Not Predictable

Figure 1.3: Changes of today.

sharp, so we cannot differentiate them mathematically. Therefore, we cannot predict the future anymore (Fig. 1.3).

1.3   BOUNDED RATIONALITY AND SATISFICING

Herbert A. Simon pointed out that rationality is bounded [1] and, outside of the bounds of rationality, we cannot optimize the result. We have no other choice than to solve the problem in a way just to get the satisfying results. Simon called this Satisficing (Satisfy + Suffice) [2].

1.4   RATIONAL WORLD TO CONTROLLABLE WORLD

SYSTEM IDENTIFICATION

When the world was small and closed, we could apply mathematical rationality in a straightforward manner. But as our world of engineering is quickly expanding with the expansion of our world, the problem came up that we cannot apply mathematical rationality in a straightforward manner as we could until then.

The word "straightforward" is used. But we should pay attention to the efforts of engineers to develop such rational approaches. Even in a small closed world, systems have large degrees of freedom. So, if we look at the system and try to solve the problem simply in a straightforward manner, the problem of computational complexity comes up and we cannot solve it. Therefore, engineers classified systems into clusters and if they find out the system belongs to the same cluster, then they can apply the model, whose degrees of freedom are reduced to the minimum, to solve the problem. This is called *System Identification* and it is nothing other than identifying this model. Then, they know what parameters they should take care of because each model has a specific set of parameters. Thus, they could solve the problems in a "*straightforward*" manner.

EXPANDING THE RATIONAL WORLD

But the world is expanding very quickly, and it becomes increasingly difficult to solve the problem this way, i.e., mathematically and get the result quantitatively. Traditional engineering required mathematical rationality and optimized results. This is, in a sense, a challenge to the bounded rationality, as Simon pointed out [1]. In traditional engineering, a strict means to an end must be established, because reproducibility played the most important role.

The path to the goal must be very clear and mathematically defined. No ambiguity is allowed, and the result must be optimum. This is a great challenge, but engineers succeeded in expanding the rational world and changing it to a Controllable World (Fig. 1.4). Then, how can they do that?

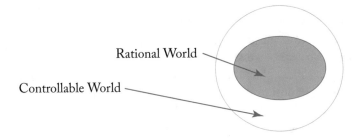

Figure 1.4: Rational World to Controllable World.

ARC FOR EXAMPLE

Let me explain by using arc as an example (Fig. 1.5).

Figure 1.5: Arc.

There is tremendous amount of research on arc. But we cannot predict its behavior. If we could, then we could prevent thunder and lightning. Still, arc is widely used in welding. Without arc welding, we cannot build buildings, bridges, automobiles, etc. Arc welding is used everywhere. Then, why can we use arc for welding?

IDENTIFICATION

It is because we do not apply rational approaches to the arc itself. We look for parts around the molten pool. These parts can be analyzed and controlled mathematically. So, if we can control these surrounding parts, we can use arc for welding.

Arc changes its state from gas to liquid and liquid to solid. So, if we pay attention to the molten pool itself, we cannot solve the problem mathematically. There is no single equation which governs all these states. Therefore, we abandon the idea of controlling the molten pool itself. Instead, we pay attention to its surrounding area. The surrounding areas remain solid all the time, so if we observe how they change or how they behave, we can apply mathematical approaches and we can achieve the optimum result.

This is the same idea as to how we identify the name of a river. If we look at the flow, its behavior changes from moment to moment. Therefore, we are unable to identify it. But, if we look around and look at the mountains, trees, etc., which do not change appreciably, then, we can identify the river by name (Fig. 1.6).

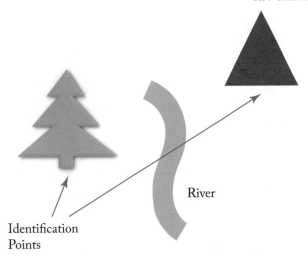

Figure 1.6: Identification of the name of a river.

The approach we took to control the arc for welding is basically the same and in this way we succeeded in substantially expanding our rational world and establishing a Controllable World.

1.5   SATISFICING WORLD

Our world is expanding so quickly that we cannot apply this approach anymore. Now, satisficing is the only solution. Even though Simon [2] proposed the idea of satisficing, he did not tell us how to do it.

In this book, we first study value and rationality in other areas. As there are many different definitions and approaches, the hope is that they will indicate to us how we could possibly achieve the Satisficing World (Fig. 1.7).

1.6   REAL-WORLD PROBLEMS—THEIR DIFFICULTIES

Artificial Intelligence (AI) is getting wide attention these days. However current AI is effective only for the world of complete information. AI beats humans, but currently it is only in the world of complete information game. Although information is abundant, it is completely structured. It is difficult for humans to search for the best (optimum) solution. However, it is easy for a computer.

We must bear in mind that our real world is incomplete and uncertain. Our knowledge is limited and our time to process a problem is limited. In the world of a game, constraints are hard; we cannot change them. But in our real world, constraints are soft and negotiable; we can

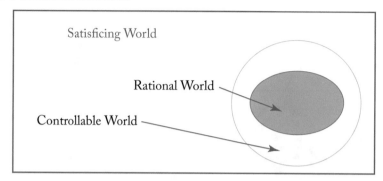

Figure 1.7: Rational World vs. Controllable World, vs. Satisficing World.

manage them if needed. Thus, other definitions of value and rationality must be studied to solve these problems in the real world.

ADAPTIVE APPROACH

Gerd Gigerenzer proposed an adaptive approach [3] for solving real-world problems. He takes up the scenario of carrying a person to an emergency hospital. In this scenario, we cannot carry out all medical testing as we do not have time. But, we must decide which emergency hospital or which medical department we should carry him/her to. Thus, we have no other choice than to make the best decision at the moment, and then apply the necessary medical treatment as time goes on. This approach is one of trial and error.

In fact, in our real world where information is incomplete and uncertain, we have no other choice than to use trial and error. That is our wisdom for adapting to the changing environments and situations.

ABDUCTION AND PDSA

Gigerenzer's approach is the same as that of Abduction proposed by Charles S. Peirce [4] and PDSA (Plan-Do-Study-Act) proposed by Walter A. Shewhart [5].

In the frequently and extensively changing world of today, adaptability has become very important. Therefore, we need to study value and rationality from many different viewpoints.

Interestingly, both Peirce and Shewhart worked in many areas, engineering being just one of them. Peirce worked in logistics, but he knew that induction and deduction do not always work in engineering and he was convinced that another logic, the Third Logic, is needed. Shewhart worked in statistics, but he also knew data-driven statistics does not always work in engineering.

PRAGMATISM

Both Peirce and Shewhart believed that a trial and error approach is strongly called for to solve engineering problems in the real world. This conviction led them to develop such a pragmatic approach. *Pragmatism* does not care how you get to the goal. Whether or not you can get to the goal that is important.

1.7 SCIENCE IS ONE PRINCIPLE, BUT ENGINEERING IS ANOTHER

Sometimes, science and engineering are regarded as the same, or more often engineering is seen as just another application of science. But this is totally wrong.

SCIENCE

Scientists search for the truth. Their goal does not change. They may seem to go by trial and error, but their final goal does not change. These trials and errors are carried out with a long-term strategy. They have nothing to do with adaptation. In other words, they look for a reproducible approach.

ENGINEERING

Engineers work to meet the expectations of their customers. They are usually very short-term and in the rapidly and extensively changing world of today, adaptability is evaluated more than reproducibility. We must keep this in mind to establish a new framework for value and rationality.

However, the wall between science and engineering is quickly disappearing.

1.8 NEW "SELFISH"

Brain science is now introducing the very interesting idea of the "Selfish Brain" [6]. In our daily conversation, "selfish" is defined as "someone who is selfish only thinks of their own advantage" [7].

But the word "selfish" in brain science is completely different. In short, it means to establish win-win relations with others. Brain scientists say if we stay "selfish" in our usual sense, we cannot survive, if the outside world changes. We may be able to make *progress*, but to survive in such a widely and frequently changing world, we need to *evolve*.

Thus, brains do not care how we, ourselves, may change. We may lose our hands, legs, etc., and transform into another completely different shape. All living things which could achieve such transformation, or another metamorphosis, are able to evolve and survive.

So, even brain scientists tell us how *adaptability* is important to survive.

Our traditional scientists and engineers may be stuck too much with the idea of reproducibility. The world around us is changing all the time so we need to *adapt* to survive. *Adaptability* might be more essential than reproducibility.

IN AI TERMS

To explain this using AI terms, our daily life definition of "selfish" is exploitation. We stay on the same track. We accumulate knowledge and experience. So, we can make *progress*, but if the outside world changes, the goal will not be effective anymore.

The other approach is exploration. The original definition of the word "exploration" is to explore how we can survive with current resources (limited knowledge, experience, etc.). In short, we explore how we can adapt to the changing outside world. So, how we can establish win-win relations with the outside world is exploration. This to brain scientists is "*selfish.*"

We have discussed how the outside world is changing. Accordingly, our engineering world is changing rapidly as well.

1.9   INDIVIDUAL PLAY TO TEAM PLAY

As the outside situations did not change appreciably and we were living in a small world, our traditional engineering focused its attention on individual products. But, to cope with the widely and frequently changing outside world, we engineers are forced to work together as a team. In addition, our products must also work together.

IOT (INTERNET OF THINGS)

The Internet of Things (IoT) is gaining wide attention. IoT is expected to establish the Connected Society. But we should be careful. It is not IoT which leads us to the Connected Society. On the contrary.

Kevin Ashton was working at Procter and Gamble in the field of supply chain management. He was taking care of RFID (Radio Frequency Identification). The quickly diversifying outside world made him aware of the needs of the Connected Society. He realized that we need a wide variety of products to work together to cope with the diversification and personalization. We cannot survive if we remain in the traditional framework of individual products [8].

THE CONNECTED SOCIETY

Thus, this brings us to the Connected Society, where not only humans but products also are connected. To express it another way, our world has changed from *individual play* to *team play*. And the members of this team are products and humans. We should not forget that although IoT is advocated to connect *things*, the word "things" include humans as well.

THE DIFFERENCES BETWEEN INDIVIDUAL PLAY AND TEAM PLAY

Another important point we should remember is that individual play and team play are completely different. Knute Rockne told us "Best team cannot be made by 11 best players. The best team can be realized by players who play best to respond to the changing expectations" [9].

Therefore, we must get out of the traditional framework of individual-product-based design and production and should move toward establishing the new team-product-based framework.

Therefore, the Connected Society is revolutionizing our engineering. So, we must reconsider value and rationality and establish the new framework.

This book introduces many different definitions of value and rationality to realize which would best meet the requirements of team play. Then, the most promising direction for our future team-product-based engineering will be understood.

Table 1.1: How our world has changed

World	Boundaries	Changes	Prediction	Rationality	Play
Closed, Small, Mathematically Rational World	Yes	Smooth	Possible	Mathematical	Individual
Closed, Expanded, Controllable World	Yes	Smooth	Possible	Mathematical	Team
Open, Satisficing World	No	Sharp	Not possible	Value	Team

CHAPTER 2

Value and Rationality: Traditional Engineering Definition

2.1 VALUE

The most well-known definition of value in engineering is the one defined by Lawrence D. Miles [10]:

$$Value = Performance/Cost.$$

Although the numerator performance has broad meaning, it has long been interpreted as functions. This is because people wanted products which were useful in their lives. And these functions are defined by the producer, because what people wanted did not change from person to person.

As the environments and situations did not change appreciably until recently, what people needed was the same or very similar kinds of products. Thus, mass production became prevalent and the producer took the initiative.

Note that cost is the denominator. In engineering, the quantitative approach is preferred. So, if the numerator performance can be defined as functions, then we can evaluate cost in manufacturing quantitatively. Thus, in addition to upgrading and increasing functions, cost reduction is called for to increase value. Value defined this way, therefore, can be evaluated quantitatively.

The rational approach permitted such quantitative evaluation. As Miles' definition of value is quantitative, it is also used widely in economics.

Rationality in our traditional engineering will be explained next.

2.2 RATIONALITY

SYSTEM IDENTIFICATION

Engineers prefer quantitative evaluation. But systems have a large number of degrees of freedom. So, as described in Chapter 1, engineers paid tremendous attention to reducing it to the minimum. Thus, they succeeded in reducing the number of degrees of freedom and in classifying them into system group categories.

Thus, what engineers needed was to identify which group category the system belongs to (System Identification). Then, they will know what parameters they must take care of. Therefore, functions and cost can be evaluated quantitatively and such rationality is mathematical rationality.

BOUNDED RATIONALITY AND SATISFICING

But Herbert Simon, Nobel Laurate in economics, pointed out that such rationality is bounded [1]. Increasing size and complexity of the system causes computational complexity and we cannot apply mathematical rational approaches. He pointed out that optimization is possible only when mathematical rationality can be applied. Beyond the bounds of rationality, we have no other choice than to settle for satisfaction. So, he proposed Satisficing (Satisfy + Suffice) [2].

INDIVIDUAL PRODUCTS

Why value can be evaluated quantitatively, and optimum result (value) can be obtained in traditional engineering, is because engineers used to work on individual products. The systems they worked on are not too large or complex. Therefore, they can reduce the number of degrees of freedom and solve the problem along the line of mathematical rationality.

GLOBAL OPTIMIZATION—ANOTHER SATISFICING

We must remember that even in the field of optimization, when it comes to global optimization, we are not really optimizing mathematically. What we do is nothing other than satisficing (Fig. 2.1)

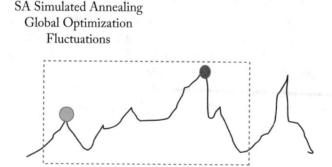

Figure 2.1: Global optimization.

For example, simulated annealing repeats searches many times. So, we believe the highest one we find after many searches is the highest (optimum). But, the higher peak might come up in the next search.

Since it only satisfies us by doing so many searches, global optimization is none other than satisficing.

CHAPTER 3

Increasing Difficulty of Recognizing Product Quality Improvement

3.1 WEBER–FECHNER LAW

Weber–Fechner proposed the following. They found out that to recognize the level of stimulus, proportional increment is needed:

$$\frac{\Delta R}{R} = constant.$$

This became known as the Weber-Fechner law and we can understand it more easily if we take up this example.

If a man who usually speaks in a small voice raises his voice, people easily understand that he raised his voice. But, if a man who always speaks in a loud voice raises his voice a little, people would not recognize.

3.2 PRODUCT SERVICE SYSTEMS (PSS)

The term Product Service Systems (PSS) is often heard of these days. But, we must be careful about what "service" here really means.

In traditional engineering, end products were our primary concern; our focus was on functions. In most cases, service in PSS is discussed in this traditional engineering mindset, i.e., service is considered as another function.

Why do engineers start paying attention to service? The following explains why. At the early stage of product development ΔS (increment of stimulus) is very large, so customers can easily recognize the improvement. However, product quality is continually improving and that quality is approaching its ceiling (Fig. 3.1).

Therefore, it becomes increasingly difficult for the producer to develop products that will convince the customer of the improvement in quality of their products. And what makes the matter worse is that the demand is rapidly diversifying and personalizing.

If service is added as another function, it will make a difference and the customer will recognize improvement in product quality easily. Thus, although the word service is used, it is discussed in the product-focused framework.

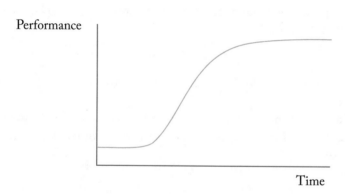

Figure 3.1: Quality improvement.

CHAPTER 4

Value and Rationality: Definitions in Other Fields

Engineers did not have any doubts about quantitative evaluation of value and mathematical rationality because they are used to such mathematical approaches. Sociologists, however, are not happy with such mathematical approaches alone. They feel there should be other definitions of value and other interpretations of rationality.

4.1 ZWECKRATIONALITAET AND WERTRATIONALITAET

Max Weber, a German sociologist, pointed out that there are two kinds of value, i.e., zweckrationalitaet and Wertrationalitaet [11]. Zweckrationalitaet is instrumental rationality or procedural rationality (Table 4.1). What is very different from the engineering definition of rationality is Wertrationaiitaet or Value Rationality.

Table 4.1: Zweckrationalitaet and Wertrationalitaet

Zweckrationalitaet	Instrumental Rationality
Wertrationalitaet	Sustantive Rationality. If expections are satisfied, MY value is secured. Therefore, all approaches to get to this goal are rational.

Weber insists that if the goal is satisfactory, the process or procedures to get to the goal does not matter. As he is a sociologist, he cites such cases as ethics, aesthetics, religion, etc. This is the same philosophy as Shakespeare's "all's well that ends well," otherwise known as Pragmatism (Fig. 4.1).

4.2 ECONOMIST'S RATIONALITY

Herbert A. Simon, American economist and political scientist and Nobel Prize Laureate, on the other hand, attaches importance to the approaches of a goal. What matters to him is whether the approach is correct and an adequate approach to the goal is established [1].

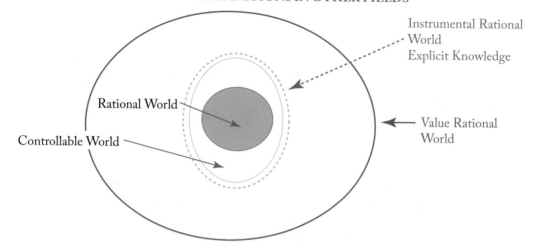

Figure 4.1: **Max Weber's value rational (Wertrationalitaet) world.**

4.3   PRAGMATISM

In the U.S., the philosophy, "if the goal is achieved, then it is fine, we do not care what path we took to get there" is called Pragmatism [12].

Although the concept of Pragmatism originated in the UK, it has been developed in the U.S. UK is a seafaring country. In fact, UK once dominated the seven seas. That is why Pragmatism originated in the UK as will be explained in Section 4.4.

We must pay attention to the interpretations of instrumental rationality or procedural rationality. Those people outside of engineering do not necessarily mean mathematical rationality. If an approach to an end is reasonable, they will consider it instrumental rationality or procedural rationality. Weber is unique because he defined another rationality, one which is based on Pragmatism.

Why Pragmatism became prevalent in the U.S. is because U.S. expanded its frontier to the West. Nobody knows what was waiting there. Yet people moved West to achieve their goals. Nobody knew how to get there, or what path to take, yet if they achieved their goal of making it there then their journey was a success. Trial and error was the only method of moving foward.

4.4   RAILROAD VS. VOYAGE

Weber's definition of instrumental rationality or procedural rationality and value rationality may be best understood if we compare them to a railroad and a voyage.

RAILROAD

In the case of a railroad (Fig. 4.2), we must select tracks to get to the destination (the goal). So, it is important to select the right track. Among the choices, you choose the fastest way or the cheapest way, whatever you like. Yes, you can optimize your journey to your goal. And you can make decisions at the beginning.

Figure 4.2: Railroad.

VOYAGE

But in the case of an ocean voyage, the situation is completely different. It may be a fine day today, but tomorrow a hurricane might attack us. In such a case, we need to drop anchor at a port we did not plan to in order to escape from the hurricane. And in the worst case, we may have to abandon the idea of getting to the port we initially planned to. Then, we have to ask ourselves why we set sail in the first place and what was our ultimate goal.

If we can achieve our ultimate goal, no matter where we may end our voyage, we are happy. We do not care what port it may be. In other words, we should ask ourselves what port will meet our expectation and achieve our ultimate goal. And we must remember that in the case of a voyage, we must make decisions continuously to respond to the frequent changes (Fig. 4.3).

4.5   ABDUCTION: WHAT MATTERS IS THE GOAL

Interesting enough, Charles S. Peirce [13], American philosopher, logician, mathematician, and scientist, proposed a very different logic, *Abduction*, which is often called *The Third Logic*. He is known as the Father of Pragmatism and is one of the famous three founders of this philosophy. The other two founders, William James and John Dewey, were psychologists. But Peirce is logician, mathematician and scientist.

Figure 4.3: Voyage.

INDUCTION AND DEDUCTION

Induction and deduction are well known as logics. We accumulate experiences and structure them into knowledge (induction). Once the knowledge is established, we can solve the problem based on it (deduction). It may not be a quantitative approach as Simon's rationality, but it is reasonable enough to be called rational (Fig. 4.4).

ABDUCTION

But Peirce's Abduction is based on trial and error. He does not care how reasonable the process is. What matters to him is if the problem is solved or not, or the goal is achieved or not. The path to get there does not matter. So, this is another very different framework of logic. His approach proceeds as shown in Fig. 4.5.

First, you "abduct" some approach, which seems to work. Then you apply it to the current problem and observe. If it works, then it is fine. If not, then look for another one. Repeat this cycle until a satisfactory result is found.

This does not agree with the other two well-known logics, induction and deduction. The other two focus their attention on how reasonably the process or the approach is carried out. Abduction does not. It focuses only on the goal.

Abduction may not sound reasonable, but if we do not know the path to the goal, there is no other way, and in this sense, this *Abduction* approach is very much reasonable.

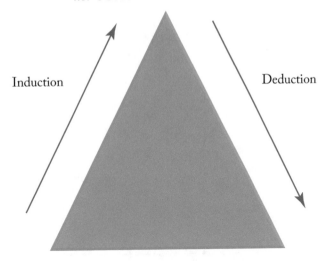

Figure 4.4: Induction and deduction.

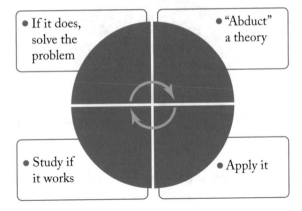

Figure 4.5: Abduction.

4.6 CONCEIVE-DESIGN-IMPLEMENT-OPERATE (CDIO)

MIT introduced CDIO (Conceive-Design-Implement-Operate). The basic idea is the same as *Abudction* and this approach is very much pragmatic (Fig. 4.6).

In other words, Induction and Deduction are tactics-focused logic, while *Abduction* is strategy-focused.

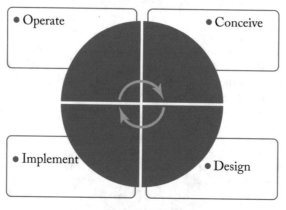

Figure 4.6: CDIO (Conceive-Design-Implement-Operate).

CHAPTER 5

Design–Another Form of Decision Making

Design is an activity of decision making. Or it would be better to say decision making itself is design. Here are some approaches focused on decision making. They are efforts in a search for adequate rationality and for securing expected value.

5.1 PLAN-DO-STUDY-ACT (PDSA)

Walter A. Shewhart, American physicist, engineer, and statistician, proposed the PDSA (Plan-Do-Study-Act) cycle (Fig. 5.1).

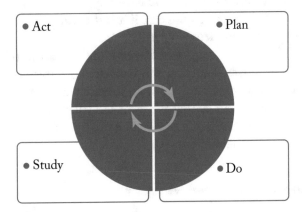

Figure 5.1: PDSA cycle.

This idea is basically the same as Abduction, the only difference being that PDSA is statistics-based. First, think of a hypothesis (Plan) and apply it to the current problem (Do), and observe if it works (Study) and if it does, then Act. If it does not, repeat the cycle. PDSA is such a tool for decision making.

But in Japan, where tactics are dominating, it was misunderstood and it was called PDCA (Plan-Do-Check-Act). Then, it is nothing more than another analysis tool. You apply your model and if it does not work, improve (Kaizen) it by checking what's wrong.

PDCA has nothing to do with design. Design is an act of decision making and it is very much strategic. But PDCA is just another analysis tool for Kaizen. Shewhart strongly opposed the use of the "word" PDCA as he insisted that it was not what he proposed.

DATA-DRIVEN AND HYPOTHESIS-DRIVEN STATISTICS

Statistics is used in two ways. One is data-driven, and the other is hypothesis-driven.

A data-driven approach extracts knowledge from data. We do not possess any hypotheses. What we get at the end is all what data provides.

Hypothesis-driven, on the other hand, is something like a (medical) diagnosis. You determine which disease explains a patient's symptoms, or illness. Because we cannot carry out all testing to find out the real cause of illness, we hypothesize what illness would exhibit such symptoms and by trial and error we discover the seemingly right cause. We should be careful that our final decision may not be a real cause. But what is important is that we find a treatment that works.

DESIGN OF EXPERIMENTS

Design of Experiments is hypothesis-driven. We have a hypothesis and we apply it to the problem and if it works, then it is fine. We do not care if it is the real cause of the problem. What is important is whether or not the problem is solved.

This is another forward reasoning. We do not know the goal. This holds true both in data-driven and hypothesis-driven statistics. But data-driven statistics solves the problem in a straightforward manner, if a way to an end is legitimately chosen. However, in Design of Experiments the search is carried out by trial and error and what we obtain at the end might not be the true solution. But if it works then it proves that the hypothesis is legitimate and we can solve the problem substantially (Table 5.1).

Table 5.1: Data-driven and Hypothesis-driven statistics

Data-driven	No hypothesis. Conclude from data.
Hypothesis-driven	Verify hypothesis.

Simon pointed out that rationality is bounded, and beyond the bounds of rationality we cannot optimize the solution. Therefore, we have no other choice than to look for the result, which satisfies us enough. But, again, Simon did not tell us how.

5.2   ADAPTIVE APPROACH: IMPORTANCE OF HEURISTICS

Gerd Gigerenzer, German psychologist, agrees with Simon's concept of the bounded rationality. He further points out [3] that as our world is complex and uncertain, we must make decisions with limited knowledge, resources, and time. But, most rational approaches ignore these constraints and they assume that we have perfect information and unlimited time.

As a typical example of such an incomplete information case, he takes up the case of carrying a patient to the emergency hospital. We cannot perform complete testing to find out which department we should bring him to. We must make a decision at once, but our information is very limited. This is the constraint in the real world. So, he pointed out the importance of an adaptive approach, which is none other than a trial and error approach such as Abduction, PDSA, etc.

Putting it another way, we do DESIGN always, because our knowledge is very much limited. But to go on, we must make a decision, no matter how limited our knowledge may be.

5.3   CURRENT AI: ITS LIMITATIONS

In fact, this is the issue AI is facing today to develop further. Current AI works so well in the complete information world, but such disparate problems as Gigerenzer took up cannot be solved by today's AI because information in our real world is incomplete.

Therefore, it must be added that real constraints may change, or we can change the constraints, if needed. Rational approaches assume that constraints are hard, i.e., they cannot be changed. But in the real world, constraints are soft and negotiable.

5.4   RATIONALITY: MATHEMATICAL AND REAL WORLD

Thus, mathematical rationality and rationality of the real world are different.

Yesterday, our world was small and closed, and changes were small and smooth. Therefore, we could mathematically differentiate them and predict the future. But today, changes are sharp, so we cannot differentiate them and predict the future.

5.5   DESIGN: ANOTHER AREA OF RATIONALITY

Anyway, we should stress that design is another area of rationality. Strangely enough, in design creativity comes up as the most important keyword, but decision making does not. But to be creative, we must decide which way we should go. We do not know anything about the world toward which we are heading. We are exploring a New World.

In the Westward Movement in the U.S., the explorers had to create their paths to achieve their goals. It is a strategic action and decisions are needed to make another step forward. Designers and design engineers should discuss more on the issue of decision making. It is a typical action of creativity.

5.6 PATTERN APPROACH

Carole Bouchard, French professor at Arts et Metiers, ParisTech, is pursuing design creativity from the standpoint of decision making, although they do not call their research as the one for decision-making for design. Their research is very interesting.

Images play an important role in design. This is something on which everyone can agree. However, images are not well organized for design. Researchers at Arts et Metiers, ParisTech introduced patterns to organize them. Patterns are used often in image processing, but what is unique with their research is that their aim is not to classify or identify images, but for decision making. If we compare patterns, we know which part of the pattern still needs to be explored. Therefore, we can find a new direction.

Although they are not doing research in brain science right now, such pattern processing may be what our brains are doing for decision making. It is expected that such a pattern-based approach to creative design will be taken up in the field of brain science very soon. And it should also be added AI will work effectively in such an approach.

The reason why Arts et Metiers, ParisTech introduced pattern-based approach may be because it is originally an engineering school, not an art school. So, to them, such a pattern approach may be more familiar than other approaches discussed in traditional art schools.

5.7 EFFECTIVENESS OF THE HYPOTHESIS-DRIVEN APPROACH

In design or in engineering we have goals on our minds. But we should remember that we cannot find our way by backward or goal-driven reasoning. Again, information in our real world is incomplete and uncertain. Therefore, we generate a hypothesis which seems to work and apply it and study if it really works or not. If it works, then it is fine. It might not be the best (optimum) solution, but regardless we can get to the goal. If it does not work, then we generate another hypothesis. Repeat this cycle until we can get to the goal. So, in most cases, design or engineering is an activity of trial and error, because they are creating a new world.

The reason why there weren't any doubts about mathematical rationality is because our world was closed and small. Thus, the problems these days were straightforwardly solved by mathematical rationality.

In other words, art schools are focusing too much on tactics. Design is a strategic activity, so we should re-consider creativity in design from the standpoint of strategy.

CHAPTER 6

Importance of "Self"

We have seen many different definitions of value and rationality. But most of them are in a sense "objective." We should consider "self" more.

6.1 SELF-DETERMINATION THEORY (SDT)

In this sense, Edward L. Deci and Robert M. Ryan's theory of Self-Determination (Self-Determination Theory, SDT) [14, 15] is very interesting.

They discovered that if we are motivated to do the job and we make a decision to do that, we feel very much satisfied. We do not care how much we will be paid. On the contrary, if we are told to do the job, we do not feel happy, no matter how much money will be paid. Thus, they made clear the importance of intrinsic motivation and self-determination.

LEGO

Lego is taken up as an example to demonstrate that processes are yielding value and process values are often more important than product values. Why people buy Lego is because they can use blocks to build what is in their imagination. They have their motivations and they can make decisions themselves. So, Lego is a typical example of how SDT is effective in business. In fact, Danish people put importance on emotional aspects in design, as we will see in the chair case introduced later.

FLOWER ARRANGEMENT

Let us take another example. In Japan, flower arrangement is a highly respected skill and there are many schools which teach this subject. Why do people pay high tuition fees to learn flower arrangement? If they hired an expert it would cost much less and it would be arranged in a much shorter time. But people enjoy learning. Why? Because they can create what is in their imagination. This is another good example of SDT.

MOUNTAIN CLIMBERS

Mountain climbers are another example of SDT. They select difficult routes to get to the top of a mountain. The more difficult the route is, the happier and more satisfied the climbers are because they can feel the sense of accomplishment and can demonstrate how capable they are.

This is an example of Maslow's self-actualization, but it is also a good example of SDT.

ABDUCTION AND PDSA

Note that Abduction and PDSA are also good examples of SDT. Which hypothesis to use is nothing other than decision making. So, when it is found out that our decision was correct, it brings more joy and satisfaction.

WHY ARE USERS CALLED CUSTOMERS?

Traditional engineering has been focused primarily on products. So, engineers have been trying to produce better quality products. Engineers are accustomed to the idea that the producer provides products to the user. But users are not users. As the fact that they are called customers indicates, they would like to customize their products.

6.2   HIERARCHY OF HUMAN NEEDS

Abraham H. Maslow proposed the Hierarchy of Human Needs [16] (Fig. 6.1).

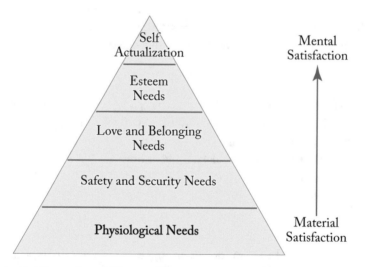

Figure 6.1: Maslow's Hierarchy of Human Needs.

At the bottom of the pyramid people want to satisfy their physiological needs. They need food, water, etc., to survive. Once these basic needs are satisfied, people would like to spend their life in safety. They would like to have housing, etc. to accommodate them in safety. When such safe living is secured, people go one step higher and they would like to have friends. They would like to belong to a community, etc., to interact with other people. When this need is satisfied, they look for esteem. They would like to be awarded. They would like to be recognized by society. Finally, at the top of the pyramid, people would like to actualize themselves. They would like to demonstrate how capable they are.

PHYSICAL SATISFACTION TO MENTAL SATISFACTION

At the lower level, humans look for physical satisfaction. They need products in their life. As is well known, the primary sector of the economy is agriculture, fishing, etc. People retrieve raw materials from nature to survive. The secondary sector is the transformation of these materials into products. The first and second sector industries were developed to satisfy the basic human needs or the lower level needs in Maslow's hierarchy.

HUMAN NEEDS TO GROW

But as Maslow, and Deci and Ryan point out, humans need to grow, so they move up the pyramid. Thus, their needs gradually change from physical to mental. This change brought the third sector where services play a leading role.

CHAPTER 7

Increasing Importance of Process Value

7.1 LEGO

Consider the Lego company (Fig. 7.1). They only sell blocks; cost is minimal. But customers enjoy combining them together in a wide variety of ways. They pay to enjoy the process. The process of combining blocks into different products excite customers. Thus, Lego demonstrates how important processes are in creating value, just as in the case of mountain climbing. Challenge is the core and mainspring of all human activities.

Figure 7.1: Lego.

7.2 CREATIVE CUSTOMERS

Figures 7.2 and 7.3 show how creative our customers are and how they would like to customize products.

Next time, going backward

Kids are very creative

Figure 7.2: Creative children.

Oh no!
You got holes there!

It's the fashion, Grandma.

Figure 7.3: Creative youngsters.

CREATIVE YOUNGSTERS

Children invent many new ways to slide down. Such invention really gives them joy and when successful, they feel the sense of achievement. Indeed, this is another good example of SDT (Fig. 7.2).

Youngsters introduce holes into their jeans. These jeans are new and of course without a hole. Why do they do that? They would like to share stories with their friends. They would like to customize their jeans. Quality or value in traditional engineering is not important (Fig. 7.3).

SELLING CHAIRS WITH FLAWS?

Denmark is well known for its design of household furniture. Fritz Hansen is one of them. What is interesting is that they sell chairs with natural leathers with a much higher price, although these natural leathers have scratches and flaws. They can produce better-quality artificial leathers. But chairs with these splendid artificial leathers are sold at a much lower price. This is because if there are scratches or flaws in natural leathers, then customers can imagine the animal's life and can enjoy sharing stories about them.

This example very well illustrates how important satisfaction is. Engineers should pay more attention to what satisfies customers. It is very much reasonable to charge a higher price for chairs with natural leathers with scratches or flaws, if we know why. Thus, value or rationality should be studied from another perspective or in a much broader perspective.

7.3   FLOWER ARRANGEMENT: KA-*DO*

In Japan, flower arrangement is very well accepted (Fig. 7.4). But it should be stressed that although it is sometimes called art, the Japanese learn to enjoy self-actualization and SDT.

Figure 7.4: Flower arrangement.

Flowers arranged by these learners may not be so beautiful as experts do, but learners feel the sense of accomplishment and enjoy the process. In fact, processes count more than the final products. Flower arrangement started in the 7th century, but at that time it was more ceremony than enjoyment. The Japanese started to enjoy the process of flower arrangement in the 16th century.

Flower arrangement is called by another name, Ka-*Do*. Ka means flowers. *Do* means the process or the way of learning. The Japanese love *Do* very much and have three typical arts of refinement: Flower Arrangement (Ka-*Do*), Incense Appreciation (Ko-*Do*), and Tea Ceremony (Sa-*Do*). Although the final one bears the name ceremony, what is important for them is the process of self-learning, not the formal procedures. So, refinement here should be interpreted as process refinement or self-refinement.

Japanese love the act of learning. This can be easily understood if we refer to the martial arts of Ju-Do and Ken-Do, as well as the other arts mentioned above. As there are countless *Do*s in Japan, this shows how much the Japanese enjoy self-actualization and how strong their need to grow is.

CHAPTER 8

Reliability to Trust

8.1   RELIABILITY

Reliability is one of the most important components of quality, i.e., value. Until very recently, reliability was evaluated by statistics. But, it was a completely data-driven statistics. Figure 8.1 shows a bathtub curve, which constitutes the basics of reliability engineering. The bathtub curve shows the change of failure rate over time. The failure rate is defined by:

Failure rate = The number of failed products/The number of remaining products.

These numbers are counted over the fixed time interval.

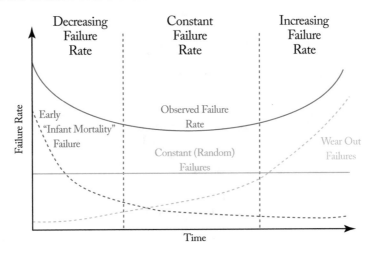

Figure 8.1: Bathtub curve.

When products were produced in mass, this approach worked very well and customers did not care too much about their own product. If their products failed, then they asked for a new one and they were happy.

8.2   TRUST

However, situations are quickly changing these days. Diversification and personalization progress rapidly, so products are becoming diversified and personalized. It therefore becomes increasingly difficult to apply such data-driven statistics approach. In addition, customers are now asking for the reliability of their own products. They say, "What I am asking about is *My Product*. I do not care what other products are."

Such reliability of an individual product is called *TRUST*. Now, customers are asking for trust, instead of traditional reliability, which reliability engineering has long pursued based on data-driven statistics. In other words, conventional reliability is very much producer-centric. Now customers are asking for customer-focused approach.

To understand the difference between reliability and trust more clearly, let us compare hardware and software development.

8.3   HARDWARE AND SOFTWARE DEVELOPMENT

HARDWARE DEVELOPMENT

Hardware is developed with fixed functions. It is developed to meet the design specifications and before shipping a prototype is produced to verify whether it really satisfies the design requirements (Fig. 8.2).

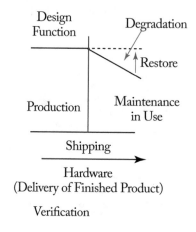

Figure 8.2: Hardware development.

We must remember the quality which the producer guarantees is the product quality at the time of shipping. The product deteriorates once it is shipped to a customer.

SOFTWARE DEVELOPMENT

In the early stages, software was developed in the same way as hardware. In the early 1960s, there were software development facilities called "software factories," which clearly indicates people did not realize the difference between hardware and software.

But, difference between hardware and software was soon realized. So, software engineers changed their system development, as shown in Fig. 8.3.

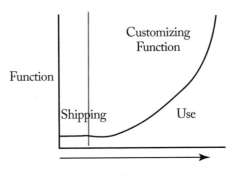

Figure 8.3: Software development.

The fundamental difference between this new software development and the old one is this new way is the development of growing functions.

Functions grow with time. Software developers introduced new prototyping approach called *continuous prototyping*. They confirm at every stage if the functions grow in response to the customer expectations. Thus, this is very much customer-focused. But what is more important is that this way of development really answered the customers' expectation of trust.

In the case of software development, first the basic functions are provided. Then, when the customers get used to them and they expect an upgraded version, software developers provide a little bit upgraded version. This step-by-step upgrading continues. Thus, it is called continuous prototyping.

What is important for the producer is when customers get used to the system and feel confident with using it, then they put trust in the system. And once when they put trust in the system, the system becomes "*MY* system" and they start asking for upgrades. They would not go to other systems which have higher functions. So, to software developers, they can secure lifetime customers and they can understand what their customers expect next. Thus, if they can satisfy such customer expectations, customers will stick to the system and are likely to grow with the system.

This way, software developers solved the problem of how they can build up trust and how they can satisfy the human needs of self-determination and growth.

Interestingly, in English, we use different words, confidence and trust. But in German, they use the same word "Vertrauen." Indeed, trust in yourself is confidence and trust in other things is trust. There is no difference.

HARDWARE AND SOFTWARE DEVELOPMENT: FROM ANOTHER ANGLE

Let us look at the difference of hardware and software development from another angle. Hardware development is carried out in an open-loop manner (Fig. 8.4).

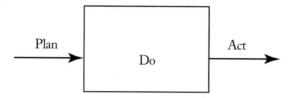

Figure 8.4: Open loop.

On the other hand, software is developed in a closed-loop manner (Fig. 8.5). In a closed-loop approach, the customer's voice is, naturally, always fed back so that customer's expectations are very well met and satisfied, and such growing function development can be carried out.

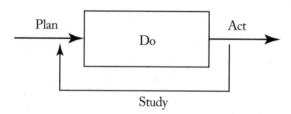

Figure 8.5: Closed loop.

REFLECTIVE PRACTICE

Donald A. Schon published the book *The Reflective Practitioner: How Professionals Think in Acton* in 1984 [17], but the idea of reflective practice has been known for a long time. In fact, it means to improve the way you work based on your own experience. So, in a way, everybody is doing this. But what is important is that the idea of reflective practice is nothing other than the closed loop approach and software developers make the most of it. In fact, reflective practice is a way for us to gain confidence. This may be interpreted as an example of how important processes are.

ECONOMIC EXPECTATIONS: SHORT-TERM AND LONG-TERM

It may interest you to know that John M. Keynes, British economist and known for Keynesian economics, teaches that short-term expectations can be made based on mathematical approaches, but when it comes to long-term, they do economic expectations on the basis of their confidence [18].

Long-term expectations fall outside of the bounds of rationality. While Simon told us satisficing is the only solution outside of the bounds, he did not show any quantitative solution [2]; Keynes, however, told us we can manage to do long-term quantitative expectation, if we feel confident.

Simon is more interested in decision making and Keynes' work is in the area of economic expectations. So, their goals are different, but it is interesting to know how confidence or trust can extend our quantitative evaluation outside of the bounds of rationality.

8.4   REPAIR: ANOTHER NEW VALUE CREATION

We have discussed how continuous prototyping software development contributes to trust building. Now, let us look at it from another angle.

If we turn the software development figure (Fig. 8.3) upside down and compare it with the hardware one (Fig. 8.2), we realize both are identical (Fig. 8.6). Then, aren't there any ways we can apply to control degradation of hardware? When hardware engineers talk about deterioration, they say they must prevent failures. But deterioration and failures are different. Of course, after deterioration failure occurs. But that is the last and final stage. Until then, deterioration goes on gradually. And if we look at deterioration from another angle, it is the process nothing other than fitting. All products come to fit us after using sometime. Then, we feel the product is MY product. It fits me perfectly. But most hardware engineers do not regard deterioration as the process of fitting. They just regard it as a step to failure. So, they believe if they can prevent failure, that is the best way to guarantee product quality. Their mindset is still stuck with mass products. But just as software is doing, if we can introduce a closed-loop control to deterioration, we could possibly reduce time to fitting and keep fitting time longer. Then, we can secure lifetime customers.

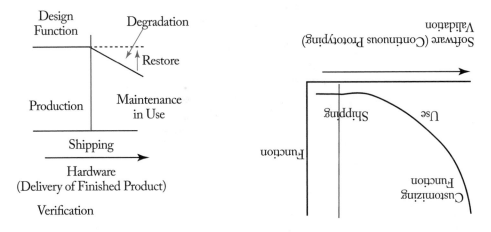

Figure 8.6: Looking at software development upside down.

Figure 8.7 shows a worn-out shoe. People or hardware people would throw this away. But ASICS, a Japanese shoemaker, has a very different view. They thought this deformation is due to the way the customers wear. If the shoes deform this way, and they produce shoes that deform this way, then they could fit the customer in much shorter time and in more comfortable way.

Figure 8.7: Wornout shoe.

So, ASICS developed shoes with the easy-to-deform middle part and the hard form-keeping part (Fig. 8.8). Although you cannot recognize any difference from the outside, people who wore this pair of shoes said "I do not feel I am putting on shoes. I feel like walking on my own feet." So, they enjoyed long walks and this product has been very successful.

Figure 8.8: Shoes for a long walk.

And, as ASICS also produces sports shoes, they applied the same idea there. Athletes do not have time to adjust, but these shoes fit their starting form immediately. So, the same idea worked very well for people to enjoy daily life as well as for sport athletes (Fig. 8.9).

Figure 8.9: Shoes for sport.

Hardware engineering should study how they can control deterioration and provide their customers with the life-long joy of best fitting (best fit to personal use). This will certainly turn their users to lifetime customers. And, it should also be added that such control of deterioration is another activity of new process value creation.

CHAPTER 9

Individual Products to Team Products–Individual Play to Team Play

9.1 WHY DO PRODUCTS NEED TO WORK AS A TEAM?

We have discussed value and rationality, but so far the discussion has focused on individual products. But, in order to cope with the frequently and extensively changing environments and situations, we need a wide variety of capabilities. What we need now is high adaptability. Then, how can we develop high adaptability?

9.2 TREE AND NETWORK

Let us consider industry and social structures. How different are they between the traditional individual product-based manufacturing and operation to team-based manufacturing?

TREE

In the 20th century, when environments and situations did not change appreciably, our strategy was fixed and did not change over time. The tree structure worked best for such a long-term single strategy, because a tree has only one output node and its structure is static, and does not change (Fig. 9.1).

NETWORK

However, as tree structure has only one node and static, it does not work in rapidly changing environments and situations. We need a wide variety of knowledge, experience, and capabilities to cope with the frequently and extensively changing situations. Thus, structures are quickly changing into a network (Fig. 9.2).

In a network, any node can be an output node. So, a network can respond flexibly to diverse situations. But to adapt to the changing situations, the network must be adaptive. To be adaptable, each node in a network needs to play a different role required to respond to the changes. Thus, each node is required to have multiple capabilities.

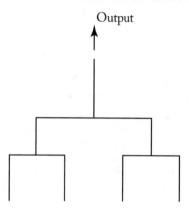

Figure 9.1: Tree.

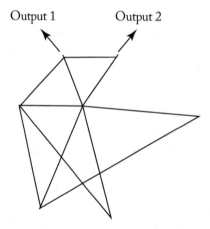

Figure 9.2: Network.

SOCCER AS AN EXAMPLE

Let us take soccer for example. In the past, the formation was fixed and each player was expected to play his role. But today games continuously change so the team is compelled to introduce an adaptive network operation. Each player is expected to play a different role from time to time which is needed to win the game. Thus, each player needs to have multiple capabilities to meet the imminent expectation (Fig. 9.3).

9.3   11 BEST, BEST 11

Knute Rockne, famous American football player and coach, has a famous quote: "11 Best, Best 11" [9]. He claims that the best team cannot be made of by the 11 best players. The best team

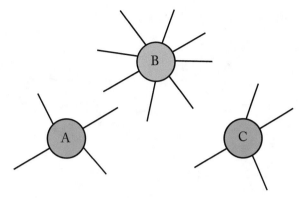

Figure 9.3: Players with multiple capabilities.

consists of the best 11 players. He demonstrated this truth by raising up the University of Notre Dame's football team to their highest ever winning level. Until then, they had been at the bottom. However, in spite of this great achievement, nobody knows the names of the players who contributed to this success. Just the team's overall accomplishments.

What is important here is not only do we need versatile players, but the strategic goal which changes from minute to minute must be understood and shared among them.

9.4   PLAYING MANAGER

In the case of soccer, Franz Beckenbauer, German football player and manager who was nicknamed Der Kaiser, introduced the Libero system, which is an adaptive network operation. He further changed the role of a midfielder. Until then, a midfielder's role is defender. But as forwards need to focus their attention toward the goal and midfielders can best see the flow of the game, he changed the role of midfielders into playing manager so, more adequate decisions can be made to turn the tide.

This teaches us a lesson that in rapidly and extensively changing situations we need a playing manager. The manager outside of the pitch is no help, because situational awareness and quick to action are requisites in strategy to respond to these rapidly changing situations (Figs. 9.4 and 9.5).

9.5   ADAPTIVE TEAM ORGANIZATION: DIFFERENCE BETWEEN A SPORT TEAM AND PRODUCT TEAM

SPORT TEAM

In such a game as soccer, the number of team members is fixed and these members do not change often. So, each member is required to have diverse capabilities to adapt to the changes that come with each game.

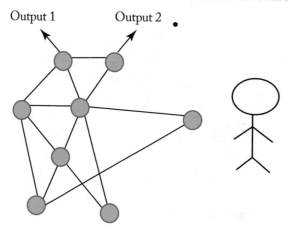

Figure 9.4: Traditional engineering (humans outside).

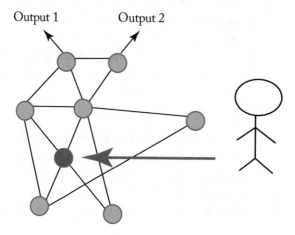

Figure 9.5: Engineering in the connected society (humans inside).

PRODUCT TEAM

But in the case of team products, situations are very different. Users may expect many different things from time to time because situations change so frequently. The organization of products they would like to use as a team varies from time to time. The number of team members and how they are organized vary to respond to customer requirements.

USER PREFERENCES

And we must consider user preferences.

Some users would like to drink cold beer so as to make the beer taste better. Others would like to keep the beer at room temperature and not make it colder.

Preference and expectation vary extensively from person to person. Thus, we need to organize a product team to meet such a wide variety of expectations. Therefore, we need to develop an adaptive team organization approach.

There is much research on team organization, but most of it deals with the issue of how we can organize a team such as a team in sports. The frequent and continuous re-organization of a team to adapt to the changes is the issue which needs immediate attention and the approach with which to tackle this problem should be developed as soon as possible.

9.6 SOCIAL NETWORKING SERVICE (SNS)

Social Networking Service (SNS) is now globalizing and it is spreading very quickly beyond national boundaries. You can easily be friends with someone in another country. But frankly speaking, SNS today is nothing more than just making friends.

The Connected Society calls for another completely different SNS. To answer customer expectation, we look for the available products across boundaries, be they national, industrial, or personal, and mobilize them.

Sharing Economy is in a way such a new form of SNS, although internet may not be used.

One of services of the new SNS would be to display sights outside the window of another car while you are elsewhere in your car. You may be driving on Route 66 in Arizona, but you can enjoy the views of Portofino, Italy which are being sent to you in real time over the Internet.

Although the outside view is not sent by the Internet in real time, the basic framework was already developed together by the Copenhagen Institute of Interaction Design and Kansei Design Group, Toyota Motor Europe (Fig. 9.6) [19]. The Core77 Design Award was given to them in 2012.

In fact, military pilots wear helmets and they do not see the real enemy plane. What they see is the image displayed. The idea is the same.

If you can equip your car with such tools that can reproduce sensory stimuli sent from somewhere far away via the Internet, you can have the smell of a sea breeze instead of the smell of cacti.

This is another sharing economy, i.e., sharing of the experience.

Figure 9.6: Window to the world [19].

CHAPTER 10

Strategy: Yesterday and Today

10.1 STRATEGY OF YESTERDAY

Early in the 20th century, changes were small and if there were any they were smooth, so that they were mathematically differentiable (Fig. 1.2). Therefore, we could foresee the future. Thus, strategy did not change. It was a long-term goal so we could focus our attention to tactics.

10.2 STRATEGY OF TODAY

Today, however, changes are so frequent and extensive. And their greatest difference from those of yesterday is that they change sharply. Therefore, we cannot differentiate (Fig. 1.3). In other words, we cannot predict what will happen next. And, what makes matters worse is the fact that situations change from moment to moment. Therefore, we are forced to make decisions every minute on how we can adapt to the changes. Thus, strategy becomes nothing more than finding a way to adapt to the changing situations.

10.3 THEIR DIFFERENCE

To put it another way, yesterday's strategy meant the goal and how it relates deeply with the tactics, but today it is more related to decision making. Strategy becomes very much short-term and ad hoc. Or, it may be better to call it Adaptive Strategy (Table 10.1).

10.4 CHANGING INTERPRETATION OF RATIONALITY

If we interpret Abduction in this way, it is indeed rational. The Oxford Dictionaries [20] defines rational as "based on or in accordance with reason or logic."

Up to now, rationality has been discussed from the standpoint of procedural rationality or tactics-focused rationality. Not too much discussion has been done from the standpoint of strategy-focused rationality.

Strategy is defined in Oxford Dictionaries [21] as "a plan of action designed to achieve a long-term or overall aim." Although the long term in this definition does not hold true any more in the frequently and extensively changing environments and situations, the part of overall aim is still true, although it may change as to how we can adapt to the changes.

What is important is that strategy is deeply associated with decision making.

Table 10.1: **Strategy: Yesterday and today**

Yesterday	Long-term. How to get to the goal.
Today	Very short-term. How to adapt to the situation.

CHAPTER 11

Modularization: Product-Based to Process-Focused

11.1 AUTOMOTIVE INDUSTRY

PASSENGER CARS

In the automotive industry, modularization is attracting wide attention these days, especially in the field of passenger cars. They are putting a large amount of effort into introducing modularization. They would like to share the common platform among different models. If they can, they can reduce cost drastically. And, indeed, they succeeded. The idea of the common platform not only reduces cost, but it also facilitates the response to the diversifying and personalizing expectations of their customers. Passenger car makers can produce common platforms in mass and produce a variety of parts so that they can prepare many models.

TRUCKS

But such an idea is already practiced in the truck industry. Truck users need different truck beds and their needs vary from user to user. But chassis, the base framework of a car, do not. So, the truck industry separated truck bed manufacturers and chassis manufacturers (Fig. 11.1).

INTERCHANGEABLE COMPONENTS

Today, modularization in the passenger car industry is progressing rapidly. They are now beginning to make their components interchangeable. Users can change components to adapt to the environment and situation and they can enjoy driving more (Fig. 11.2).

ELECTRIC VEHICLE (EV)

At this time, these interchangeable components must be changed by experts, but when the time comes that EVs become very popular, users can choose parts and buy them as they like, and they can assemble their own cars by themselves. They enjoy not only driving, but also the process of assembling parts, just as in the Lego example discussed earlier.

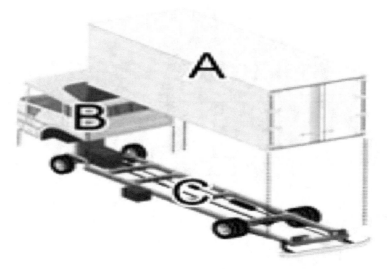

Figure 11.1: Trucks.

Figure 11.2: Interchangeable components—Daihatsu Copen.

Another important point is that not only does an EV provide such a rich experience, but it may also revolutionize the automotive industry itself. The automotive industry does not need to have a large factory anymore. Their jobs are separated into particular fields or expertise. In short, the automotive industry today is an integrated industry, but tomorrow it will be composed of a group of specialized industries.

INTEGRATED COMPANY AND SPECIALIZED COMPANY

When mass production was prevalent, integrated companies were doing good business. But in an age of diversification and personalization, they have many problems. First, it is based on producer-centric idea, so that they have a great difficulty in understanding what their customer really wants. Second, as the number of employees is huge, they need to establish a common goal among them. But the situations today call for an adaptive strategy. They do not know how they can form an adaptive team. Thus, they cannot establish brand easily.

On the other hand, specialized companies sell special products so customers understand the product and they know what to expect. Therefore, the company understands their expectations and satisfies them exactly. Thus, there is no discrepancy. So, it is easy for specialized companies to establish their brands and explore new markets easily.

NEW INDUSTRY FRAMEWORK IS EMERGING

The progress of modularization will change the industrial framework itself, as will be discussed later in Chapter 13. This transition of the industrial framework is caused by progressing modularization, but IoT introduced the Connected Society. So, this transition will be accelerated and our engineering will change drastically.

11.2 FASHION INDUSTRY
COMMON PLATFORM

The idea of a common platform was introduced in the fashion industry a long time ago. They hold fashion shows to identify which parts attract customers' attention using an eye tracker system, etc. They produce the parts which people do not care much as a common platform. And they prepare a wide variety for the parts which attracted customers' attention.

WEDDING DRESS

A wedding dress is a good example. Every lady would like to wear HER wedding dress. But not every lady is rich. So, they have to rent it. The dresses in rental shops are composed of the common platform and parts that distinguishes hers from others (Fig. 11.3)

In fact, we practice this in our daily life. Although we wear the same dress, we change accessories to suit to the environment and situation. And it also must be added that we often make accessories ourselves. Again, this is another example of process value and the wearer can be proud of herself or himself.

KIMONO

Although a Kimono, traditional Japanese garment (Fig. 11.4), belongs to the fashion industry, the modularization approach between Western dresses and a kimono is completely different. Kimonos are wrapped around the body, while Western dresses modularize the part to fit to

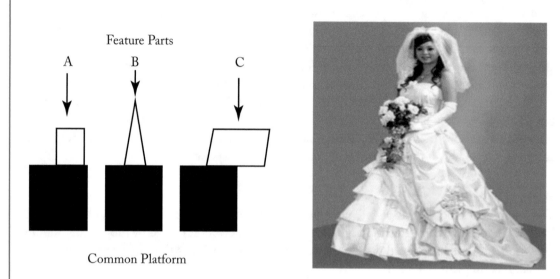

Figure 11.3: Feature parts and common platform in the fashion industry.

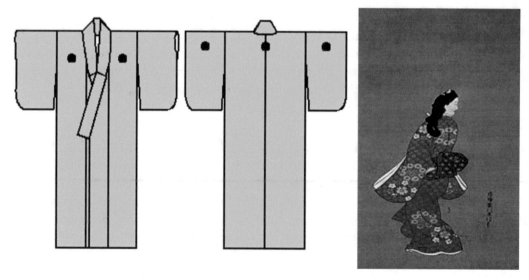

Figure 11.4: Kimono.

the body. Kimonos are made from a single bolt of fabric called *Tan*. *Tan* comes in standard dimensions—about 36 cm wide and 11.5 m long—and the entire bolt is used to make one kimono. The finished kimono consists of four main strips of fabric—two panels covering the body and two panels forming the sleeves—with additional smaller strips forming the narrow front panels and collar. Historically, kimonos were often taken apart for washing as separate panels and resewn by hand. Because the entire bolt remains in the finished garment without cutting, the kimono can be retailored easily to fit another person [22]. In fact, many kimonos are retailored. Young ladies retailor the kimonos their grandmothers wore by changing them to a modern style and wear them.

SARI

The Indian sari is well known [23]. Indian ladies drape a single bolt of fabric in a different style. They do not modularize the product as in a kimono. They modularize the processes of wearing it (Fig. 11.5).

Figure 11.5: Sari [23].

Therefore, we should take note that there are different modularization approaches. Today, most discussions about modularization focus on the Western approach. But, we should remember that there are other approaches.

11.3 BUILDINGS

HOW BUILDINGS LEARN

Stewart Brand, an American writer, best known as editor of the *Whole Earth Catalog*, published a very interesting book *How Buildings Learn: What Happens after They're Built* [24].

What he asserts is that buildings which adapt to the changes over long periods are not the ones designed and built by famous architects, but the ones with low-cost, standard design that people are familiar with, so they can modify or adapt them easily. Smart buildings may be "smart" in one sense, but from another angle, it does not meet the requirements. So as time goes on, these "smart" buildings cannot adapt to the changes. They may be smart, but they cannot *learn*. But the low-cost, standard buildings we are familiar with *learn* how to adapt to the changes. So, they survive and evolve.

Among his many illustrations of the example of *learning* buildings, Fig. 11.6 shows Plazzo Publico in Siena which survived 500 years. Although not modularized, people modified the building time after time in response to changing times. We should bear in mind value changed in response to changing times.

Figure 11.6: Plazzo Pubblico [24].

Another illustration in Brand's book is Fig. 11.7, the London Docklands. This is a typical example of how modularization is effective to adapt to the changes. We can combine modules in a different way so that they work best for the current needs.

Figure 11.7: London Docklands [24].

SHEARING LAYERS

Frank Duffy, a British architect who designs office space, proposed the idea of Shearing Layers [25]. His idea of open office space as we often see today influenced Stewart Brand.

It would be interesting to know that such open space design was very prevalent in Japan until very recently. In Japan, people work as a team, not as an individual. So traditional design was an open space. Japanese office spaces are now changing and they look as they did yesterday.

THE TIMELESS WAY OF BUILDING

About 20 years before Brand's book, Christopher W. Alexander, American architect and design theorist, published the *Timeless Way of Building* [26] and *A Pattern Language* [27]. His assertion is similar to Brand's: buildings that adapt to the changes of time are valuable and he insists that pattern is a very effective tool.

As described earlier, Carole Bouchard at Arts et Metiers, ParisTech is adopting a Pattern Approach. Her goal is focused more on how designers can respond to customer expectations, but to satisfy their diversifying and personalizing needs is also another adaptation.

JAPANESE HOUSING

Columns and Walls

House design differs between Japan and the West. In Western design, each module has its specific purpose. For example, pillars or columns support weights, and walls in between do not. Their roles are clearly distinguished. But in traditional Japanese housing (Fig. 11.8), both columns and walls support weight. There is no distinction between them. So, we can move columns freely. Therefore, we can change room layouts easily.

Figure 11.8: Traditional Japanese house.

Tatami

This is an example of continuous modularization. But a tatami, a traditional Japanese floor mat, is a typical example of discrete modularization as we are discussing today. Again, the idea of tatami is very different from that of the Western modularization.

In the West, for example, wall-to-wall carpet is highly valued. But how Japanese determine the size of tatami is very different as it is primarily based on our body movement. The half size of a tatami is the space we occupy when sitting. Its full length size is the space we occupy when we lie down. So the module size of tatami is based on how much space we occupy when we move our bodies. As walls and columns share strength equally in traditional Japanese housing, room layouts are easily changed from one layout to another. Interestingly, the size of tatami is the size one man can carry. So, we can change room layouts quite easily.

Standard Minimum Room Size

Another interesting fact is that the standard minimum room size is 4.5 tatamis. We can expand it to 6 tatamis, or 8 tatamis, or more as needed. But this minimum size is exactly the same size as Cave Automatic Virtual Environment (CAVE) (Fig. 11.9) [28]. CAVE provides immersive virtual reality environment, but its size may be closely associated with our cognition ability. The minimum size of a traditional room may be determined based on our cognitive ability.

Partitions (Shoji, Fusuma, etc.)

Another uniqueness of Japanese housing is that partitions do not separate a space.

A shoji (Fig. 11.10) is a holistic sensor. It not only conveys sounds of the outside, but also humidity, temperature, etc. So, even when we are in a room we know what is going on and how the weather is outside. Western sensors are developed to detect a particular signal. So,

Figure 11.9: **CAVE** [28].

Figure 11.10: **Shoji.**

detectability is very good with respect to this particular signal, but we cannot understand the whole environment or situation.

Fusuma are vertical rectangular panels which slide from side to side to redefine spaces within a room or to act as doors (Fig. 11.11). Western doors shut off sounds of the next room completely. But sounds come through Fusuma, so we understand what is going on in the next room. Japanese do not separate two spaces, they would like to know what is going on in the next space to be adaptive.

Figure 11.11: Fusuma and Shoji.

11.4 ORIGAMI

Most current discussion about modularization is how we can modularize final products into discrete parts to adapt to the diversifying and personalizing needs of customers.

This is a discrete modularization and the goal is to realize a product with wide adaptability. Thus, the goal is single and fixed.

ORIGAMI EXPERIENCE

But, we must remember there is another modularization. It proceeds in the opposite direction. It starts with a single paper and is folded into different shapes. Many readers may have experienced folding a paper into a crane (Fig. 11.12) or, if more experienced, then into a box (Fig. 11.13).

ORIGAMI ARTWORK

But most readers will be surprised to find out that such artwork can be folded (Figs. 11.14 and 11.15). These are artworks of Vincent Floderer, famous French origami artist.

PAPER CATHEDRAL

Shigeru Ban, Japanese architect and recipient of the Pritzker Prize 2014, the most prestigious prize in architecture [29], is known for his innovative work with paper. His paper house is not exactly origami, but he makes the best use of the idea of origami (Fig. 11.16).

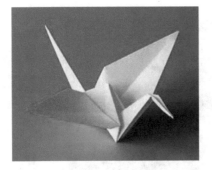

Figure 11.12: **Crane.**

Figure 11.13: **Box.**

Figure 11.14: Serpent (Vincent Floderer, French origami artist).

Figure 11.15: Tree (Vincent Floderer, French origami artist).

DEPLOYABLE STRUCTURES

Strictly speaking, deployable structures are not origami. But deployable structures [30] are designed based on the idea of origami. As they are very adaptable, they are a must in space exploration and for many applications (Fig. 11.17).

ORIGAMI ARCHITECTURES

As the idea of origami can be applied to diverse applications, many architects are now beginning to utilize the idea of origami in their design.

Figure 11.16: Paper cathedral (Shigeru Ban–The Pritzker prize winner 2014) [29].

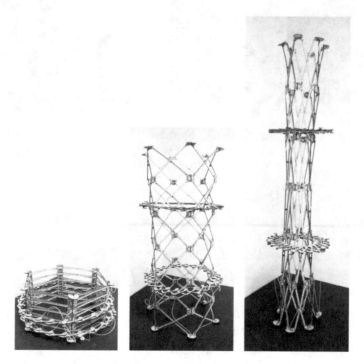

Figure 11.17: Deployable structures [30].

11.5 MATERIALS

DISCRETE MODULARIZATION AND CONTINUOUS MODULARIZATION

We have studied two different types of modularization. One is discrete modularization, whose goal it is to increase adaptability of a product. The other is in the opposite direction. Starting from an individual product to realize many different products. Thus, there are many different goals. Origami is a typical example of this second type of modularization. It may be called continuous modularization.

SEAMLESS MODULARIZATION

But there are other modularizations. One of them may be called seamless modularization. In fact, welding is one of them. In welding, two different parts are joined into one.

Strangely enough, we do not discuss composite materials from the standpoint of modularization. But it is certainly another modularization.

Yesterday, we had no other choice than to select materials which are available. In fact, materials were hard constraints in engineering design. But today, attributes can be re-organized and combined as we like. For example, Cellulose Nanofiber (CNF) is emerging [31]. This is an engineering version of "Look Back into the Future." A very promising horizon is emerging with this seamless modularization of materials.

Anyway, be it discrete or continuous, digital or analog, modularization is very effective to increase adaptability.

CHAPTER 12

Sectors of the Economy

12.1 FIVE SECTORS OF THE ECONOMY

Economists define sectors of the economy (Table 12.1).

Table 12.1: Five sectors of the economy

Sector	Activities
5. Quinary Sector	Decision making
4. Quarternary Sector	Knowledge and ICT industry
3. Tertiary Sector	Service industry
2. Secondary Sector	Transforms raw materials into products—manufacturing, etc.
1. Primary Sector	Extracts raw materials from nature—agriculture, fishing, etc.

The primary sector is agriculture, fishing, etc.; they extract raw materials or basic food from nature. The secondary sector is manufacturing, etc.; they transform raw materials into products. The tertiary sector is the service industry. These three sectors are well known. The quaternary sector is knowledge and ICT industries.

But, I don't believe many readers have heard about the quinary sector. This sector is associated with decision making. Economists point out that after the quaternary sector, decision making will play a crucial role in the economy.

12.2 QUINARY SECTOR

We have discussed earlier that there are two kinds of rationalities. One is tactics-based. The other is strategy-based. We further pointed out that strategy is quickly changing from long-term to short-term and its role is changing from providing a goal to tactics to decision making as to how we can adapt to the frequent and extensive changes.

Thus, the new quinary sector which economists proposed sounds very adequate and to the point. We are moving from a tactics-oriented strategy to adaptability-centric strategy. Thus, the role of decision making is increasing in importance.

12.3 DECISION MAKING: YESTERDAY AND TODAY

Decision making which economists take up is that of the highest level in the society. They discuss decision making in the producer-centric frame of mind and their discussion is based on a tree structure society. But, as pointed out many times earlier, society is changing rapidly from a tree structure to a network to respond to the rapidly changing situations.

Indeed, decision making becomes crucially important. But it is decision making in a network environment. And as decision is called for every minute to cope with the rapid changes. Only players know how the game is going on. Thus, decisions must be made by a playing manager on the pitch, not the manager off the pitch.

Again, we must remember decision making yesterday was to establish a goal for tactical operations, and it does not change over time. However, decision making today is to find ways to adapt to the changing situations.

CHAPTER 13

Sharing Society

13.1 SHARING ECONOMY

Sharing Economy is getting wide attention these days. Why it spreads so fast may be because people come to look at products in a much broader perspective than before. They realize that the same product can be used in a wide variety of ways.

To put it another way, no more "Grass is Greener on the Other Side of the Fence." The fence is gone. So, we can enjoy the neighbor's grass and neighbors can find another way of enjoying it.

13.2 CHANGING INDUSTRIAL FRAMEWORK

As modularization progresses, our industrial framework has changed from a open-loop, one-way system (Fig. 13.1) to a closed-loop system with feedback (Fig. 13.2). Yesterday, we, engineers, had to use available materials. And the choice is very limited. Therefore, engineers started their activities with this hard constraint. But at that time, the main stream of industry was producer-centric, so they could manage to develop good-quality products.

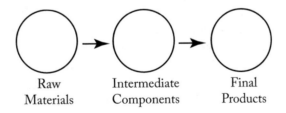

Figure 13.1: Traditional industry framework.

The progress of modularization and rapidly increasing diversification and personalization changed the scene. Now, the stream flows in the opposite direction and a customer-centric industrial framework is emerging. But, this shift also benefited engineers as well. Now, they can select materials among a very wide variety of choices. Materials are no more hard constraints. And, if engineers wish, material developers can possibly come up with a new material which answers their requests. Materials are now soft or negotiable constraints.

Not only at material level have we a wide variety of intermediate components. Until today, these component companies worked for specific integrated companies. They worked under their

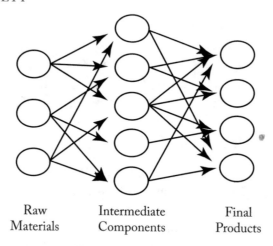

Raw　　　Intermediate　　　Final
Materials　　Components　　Products

Figure 13.2: Emerging industrial framework.

umbrella so it was not easy for them to explore new markets. But today, they do not have to work under a specific umbrella. They can look for new markets where their components can be used. Thus, market dynamism is emerging.

As this emerging industry framework is customer-centric, the best benefit goes to customers. They can expect a wide variety of products and these expectations are easily satisfied.

Sharing Economy reflects the fact that customers are now looking at products and services from many different angles. If they look at products or service from another angle, they offer another value. So, value is not unique anymore. Value varies with situations and with expectations. That is why Sharing Economy is now getting such wide attention. It symbolizes the transition from a traditional open-loop, one-way industrial framework to the closed-loop industrial framework with deep consideration of customer expectations. The emerging industry framework is nothing other than neural network or parallel distributed processing.

So, we can make full use of AI. Although the current AI is still working in the framework of complete information, there are many issues that can be processed even by the current AI. Further, such a new industry framework is expected to stimulate AI to accommodate our real world of incomplete information.

13.3   DEEPER AND DEEPER TO WIDER AND WIDER

Our traditional engineering is to stay on the same track and to make efforts to go further ahead. But in this new engineering, we stop whenever necessary and look around to find out if there are better choices. The mechanism of backtracking is what we should do in our decision making. Or, it is the way of developing adaptive strategy (Fig. 13.3). In other words, engineering has changed from deeper and deeper to wider and wider.

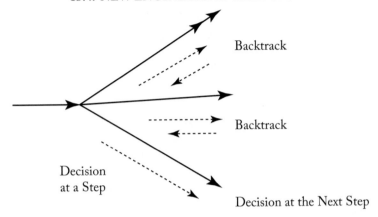

Figure 13.3: Deeper and deeper to wider and wider.

13.4 NEW ENGINEERING WILL CREATE SHARING SOCIETY

In this sense, Sharing Economy and the Connected Society, which will be described in the next chapter, have much in common. We are getting free from the restraints of performance of individual products. In short, we come to realize how versatile each product is. We should utilize their versatility. Sharing Economy is a precursor to the next stage of engineering of team products in the Connected Society. Thus, when we design a product team, we should pay attention to what functions to share and what functions to distribute to each product.

To put is another way, engineering is no more developing products or service, but it will develop a new society, i.e., a Sharing Society, where not only products, but also processes, are shared.

CHAPTER 14

The Connected Society

14.1 INTERNET OF THINGS (IOT)

Some people think IoT is technology-driven. But, in reality, it is not. It is needs-driven. Kevin Ashton, British technology pioneer, who coined the word "Internet of Things," worked at Procter and Gamble to manage supply chain. He became aware that a new approach is called for to manage the quickly progressing change of supply chain. That led him to IoT [8, 32].

IoT brought us the Connected Society, where not only are products connected, but products and humans are also connected. But it should be stressed that these things are not connected physically. Just like soccer players, they are connected by sharing information.

Until now, we have been focusing our attention on individual products and the situations did not change appreciably. So, we can manage machines or products from the outside, in the same way a soccer coach or manager gives instructions from outside the pitch.

But to grasp the quickly changing situations without delay and with high accuracy, we need a playing manager on the pitch. In other words, users need to be a playing manager.

So, the role of engineers changed. They should develop products (team members) to support them. If some products became aware that some actions are needed, they are expected to give such a signal to a user to be situational aware and let him or her take quick and adequate actions. Thus, the meaning of rationality will be widely changed. If the team can adapt to the changes adequately, then the adaptive strategy is rational. It is no more tactical rationality.

The meaning of value will also change. Until now, we discussed value with respect to individual products. We discussed performance of individual products. So, in most cases, this performance is nothing other than functions of the product. But in the Connected Society, performance is how a team adequately adapts to the minute-to-minute changes. Thus, performance tomorrow will be adaptability.

How we, engineers can support customers to adapt to the frequent and extensive changes will be evaluated and when customers feel satisfied how beautifully they could manage to adapt to the situation, it is Performance. Performance tomorrow is Adaptive Strategy.

14.2 DIFFERENCE BETWEEN THE CONNECTED SOCIETY AND SHARING SOCIETY

In the previous chapter, we discussed the Sharing Society and pointed out that industry framework will change drastically from the traditional, open-loop, producer-centric system to the neural network or parallel-distributed processing structured, customer-driven one. But even in such a new industry structure organizations do not change. It is similar to soccer. Team players do not change. What is expected from each player is what changes—they need to have versatile capabilities to respond to such expectations.

But in the case of the Connected Society, team organization changes from time to time in response to the changing needs. Sometimes, we need a group of products to work together to meet our immediate expectations, but other times we need another group, which members are largely different to do some other jobs.

14.3   ADAPTIVE ORGANIZATION

Getting straight to the point, the Connected Society will call for adaptive team organization. There is a great deal of research on team organization, but most of it discusses how we can organize a team to achieve a goal. However, situations today change continuously so that goals, i.e., what action we should take, changes from minute to minute.

It is not too much to say that there is no, or very few, if ever, research which deals with such a problem of how we can manage adaptive organization. When we say organization, it usually means a fixed number of members. But in the Connected Society, the number of team members and their goals vary from moment to moment.

Industrie 4.0, a German project, is often discussed together with IoT. But from the standpoint of team organization, they are completely different. Industrie 4.0 is based on current Small and Medium Enterprises (SMEs), i.e., the team members are fixed. In fact, the goal of Industrie 4.0 is to increase global competitiveness of SMEs. Thus, the goal is fixed, and the team members are also fixed. But in the case of the Connected Society, which IoT brought forth, the number of team members and their organization change constantly.

14.4   PROGRESS TO EVOLUTION

In other words, our world yesterday was based on *PROGRESS*. We stay on the same track and we learn how to go further ahead. But the world which the Connected Society is bringing forth is the world of *EVOLUTION*. Our engineering has been focusing on progress, but the Connected Society will change it to be evolution-focused.

14.5   TEAM FOR PROGRESS AND TEAM FOR EVOLUTION

As discussed earlier, brain scientists are now introducing a new concept of Selfishness. Their new definition is how we can establish a win-win relation with others. This is another evolution. In fact, because our brain is selfish in this sense, we could survive and keep our species through the changing environments and situations.

We must take note, however, that the core members may be the same, but the whole organization changes to adapt. That is why we have so many different kinds of living things.

Diversity is emphasized to increase adaptability. Indeed, in such an organization as a soccer team where the number of team members is fixed, we need members as diverse as possible and each member is expected to have diverse capabilities as well. But in the Connected Society, the members vary from time to time in response to user needs and each member is not necessarily expected to be multi-functional. Even if a product has only one function, we can connect other products to fulfill the functional needs of a team. In this sense, diversity does not count so much as it is discussed today. Why we emphasize diversity is because the team organization is fixed.

14.6   EXPLICIT ADAPTATION AND IMPLICIT ADAPTATION

Diversity counts when we can assume how the flow of the game changes. In soccer, there are many different game flows, but still the number of players is 11 and there are strict rules. Thus, it is another complete information game. But in the case of the Connected Society, there are no rules and the number of players varies in response to the needs.

This reminds me of the book *The Tacit Dimension* by Michael Polanyi [33]. He pointed out that there is knowledge which cannot be verbalized. For example, we cannot teach others how to ride a bicycle. It is because the situations continuously change. So, we cannot apply the traditional engineering method of system identification. In all cases, where the idea of system identification can be applied, knowledge can be expressed verbally or explicitly.

If we can verbalize how we can adapt to the change, then a fixed organization such as soccer team would work, because we can communicate with other members. And diversity would be effective as is discussed today. But in the case of the Connected Society, it is a world of tacit knowledge. We cannot communicate verbally. Or, to be more exact, we do not share the common language. Some members may be able to communicate in one language, but others do not understand it as they speak another language.

Swimming is another or better example of tacit knowledge. Although fluid dynamics progresses so much, we cannot predict the behavior of a flow. We cannot identify the parameters to control a flow. So, what we do to swim is to adapt to the current behavior of flow. Swimming is a typical example of adaptation and tacit knowledge.

In fact, most tacit knowledge is our wisdom to adapt to the environments and situations. I would say we should say wisdom instead of tacit knowledge. It is not structured and organized as knowledge. And, most people take up complexity as a great challenge today. But we must remember that complicatedness and complexity are different. There are many problems related to complicatedness. Arc is one example. Deterioration is another. There are many other complicated areas. It is our engineering wisdom to have overcome these complicated issues. Although they are not taken up as tacit knowledge, but they cannot be easily verbalized and, again, we need wisdom to overcome these complicated issues.

14.7   INSIDE OUT AND OUTSIDE IN

Western culture starts from the inside or from the self. Westerners would like to insist "self" and work toward the outside world to realize his or her dream.

But the Japanese are the complete opposite. The Japanese observe the outside world and decide how he or she would behave. As is well known, the Japanese are exceptionally homogenous in the world. So, as the proverb "The nail that sticks out gets hammered down" indicates, good results do not come from self-assertion.

If we observe which way people are going, it is best to follow them. In other words, we go with the flow. So, Japanese culture is outside in. But this is another wisdom. Westerners seem to believe they know the world inside out. When the world is closed and small, this may have worked well. But our world is not a world of complete information. It is a world of incomplete information and with uncertainties.

The Japanese way of observing first how the outside world is going would be another effective way to adapt to the frequent and extensive changes today. Engineering yesterday was explicit knowledge based, but engineering tomorrow would be moving toward the world of tacit knowledge or wisdom. It is hoped that looking back on the traditional Japanese culture will provide an opportunity in one way or another to re-examine how we can adapt to the continuous changes.

14.8   NO WALLS BETWEEN ART, SCIENCE, AND ENGINEERING

Adaptability has been discussed as a major topic in the connected Society, but we should not forget that "connecting" does not only mean to connect products, but it also means that many different fields will be connected and the walls between them are disappearing. For example, origami is used in art, architecture, and deployable structures. Other researchers are pursuing mathematical approaches.

CHAPTER 15

New Horizons Are Emerging

Thus, New Horizons of Value Rational Engineering stretch out in front of us. We can explore them as we like.

Value in this emerging engineering is nothing other than satisfying our expectations. Expectations vary from person to person and when we feel satisfied enough, we feel the product, or the process is, valuable. It should be emphasized that the value of processes is rapidly increasing. It is more related to human needs.

In other words, the emerging New World is customer-centric. Our Old World has been too much producer-centric.

When our expectations are satisfied, we do not care how they are satisfied. All's Well that Ends Well. If My Value can be secured, all approaches to this goal are reasonable, i.e., rational.

Let us make a big step forward toward the New World of Value Rational Engineering. I wish you all great success!

References

[1] https://en.wikipedia.org/wiki/Bounded_rationality 4, 5, 14, 19

[2] Herbert A. Simon (1947). *Administrative Behavior: A Study of Decision-Making Processes in Administrative Organization*, New York, Macmillan. viii, 4, 7, 14, 41

[3] Gerd Gigerenzer and Reinhard Selten, Eds., (2002). *Bounded Rationality: The Adaptive Toolbox*, Cambridge, MIT Press. 8, 27

[4] https://plato.stanford.edu/entries/peirce/ 8

[5] http://www.cologic.nu/files/evolution%20of%20the%20pdsa%20cyclel.pdf 8

[6] Yoshiaki Kikuchi et al. (2018). The selfish brain: What matters is my body, not yours?, Shuichi Fukuda Ed., *Emotional Engineering*, vol.6, pp. 49–61, London, Springer. DOI: 10.1007/978-3-319-11555-9. 9

[7] https://dictionary.cambridge.org/dictionary/english/selfish 9

[8] Kevin Ashton (2009). That internet of things thing, *RFID Journal*, June. 10, 73

[9] https://mindgames.gg/knute-rockne-on-teamwork-in-sports/ 11, 46

[10] Lawrence D. Miles (2015) (Kindle Edition). *Techniques of Value Analysis and Engineering*, Portland, Lawrence D. Miles Value Foundation. 13

[11] Max Weber (1978). *Economy and Society*, Oakland, University of California Press. DOI: 10.7312/blau17412-056. 19

[12] https://plato.stanford.edu/entries/pragmatism/ 20

[13] https://en.wikipedia.org/wiki/Charles_Sanders_Peirce 21

[14] Edward L. Deci and Robert M. Ryan (1985). *Intrinsic Motivation and Self-Determination in Human Behavior*, New York, Plenum. DOI: 10.1007/978-1-4899-2271-7. 29

[15] Robert M. Ryan and Edward L. Deci (2000). Self-determination theory and the facilitation of intrinsic motivation, social development, and well-being, *American Psychologist*, 55, pp. 68–78. DOI: 10.1037//0003-066x.55.1.68. 29

[16] Abraham H. Maslow (1943). A theory of human motivation, *Psychological Review*, vol.50, no.4, pp. 370–396. DOI: 10.1037/h0054346. 30

[17] Donald A. Schon (1984). *The Reflective Practitioner: How Professionals Think in Action*, New York, Basic Books. 40

[18] John M. Keynes (2016) (Kindle Edition). *The General Theory of Employment, Interest and Money*, Seattle, Stellar Classics. DOI: 10.1007/978-3-319-70344-2. 40

[19] https://www.core77.com/posts/23090/Core-Design-Awards-2012-Window-to-the-World-Professional-Winner-for-Speculative-Design 49, 50

[20] https://en.oxforddictionaries.com/definition/rational 51

[21] https://en.oxforddictionaries.com/definition/strategy 51

[22] Liza Dalby (2001). *Kimono: Fashioning Culture*, Seattle, University of Washington Press. 57

[23] https://en.wiktionary.org/wiki/sari 57

[24] Stewart Brand (1994). *How Buildings Learn: What Happens After They're Built*, New York, Viking Press. 58, 59

[25] https://en.wikipedia.org/wiki/Shearing_layers 59

[26] Christopher W. Alexander (1979). *The Timeless Way of Building*, Oxford, Oxford University Press. 59

[27] Christopher W. Alexander (1977). *A Pattern Language: Towns, Buildings, Construction*, Oxford, Oxford University Press. 59

[28] https://en.wikipedia.org/wiki/Cave_automatic_virtual_environment 60, 61

[29] https://en.wikipedia.org/wiki/Shigeru_Ban 62, 64

[30] https://en.wikipedia.org/wiki/Deployable_structure 63, 64

[31] https://en.wikipedia.org/wiki/Nanocellulose 65

[32] https://en.wikipedia.org/wiki/Kevin_Ashton 73

[33] Michael Polanyi (2009) (Reissue Edition). *The Tacit Dimension*, Chicago, University of Chicago Press. DOI: 10.1016/b978-0-7506-9718-7.50010-x. 75

Author's Biography

SHUICHI FUKUDA

Shuichi Fukuda is Advisor to the System Design and Management Research Institute of Keio University, Japan and Professor Emeritus of Tokyo Metropolitan Institute of Technology and Tokyo Metropolitan University. He was a Research Associate at University of Tokyo, Associate Professor at Osaka University, Japan, Professor, Dean of Engineering and Dean of Library and Information Systems, Tokyo Metropolitan Institute of Japan, and Director of Government, Industry and University Collaboration Center, Tokyo Metropolitan Government. He was a Consulting Professor, and Visiting Professor at Stanford University, a Visiting Professor at West Virginia University, Cranfield University, UK, Osaka University, Japan, The Open University, Japan, and a Visiting Associate Professor at University of Tokyo, Japan. He is the editor of many engineering books, in addition to his own numerous research publications. He served as President of ISPE (International Society for Productivity Enhancement), Vice President of IEEE Reliability Society, Deputy Group Leader of Systems and Design, ASME, and Division Chair of Computers and Information Engineering, ASME. He is a Member of the Engineering Academy of Japan, Honorary Member of JSME, Glory Member of REAJ, and a Fellow of ASME, IEICE, and ISPE.

Printed in the United States
by Baker & Taylor Publisher Services